YOUR KNOWLEDGE HAS VALUE

- We will publish your bachelor's and master's thesis, essays and papers

- Your own eBook and book - sold worldwide in all relevant shops

- Earn money with each sale

Upload your text at www.GRIN.com and publish for free

Omar Amoretti

VoLTE Capabilities. RAN Optimization Requirements

GRIN Publishing

Bibliographic information published by the German National Library:

The German National Library lists this publication in the National Bibliography;
detailed bibliographic data are available on the Internet at http://dnb.dnb.de .

Imprint:

Copyright © 2015 GRIN Verlag, Open Publishing GmbH
Print and binding: Books on Demand GmbH, Norderstedt Germany
ISBN: 978-3-668-00091-9

This book at GRIN:

http://www.grin.com/en/e-book/300821/volte-capabilities-ran-optimization-requi-
rements

VoLTE capabilities

RAN optimization requirements

By: Omar AMORETTI CASANA, BSc

Vienna, March 2015

Abstract

With LTE a relatively flat, all-IP access technology becomes feasible along with a bandwidth-efficient method of carrying multiple types of user traffic at the same time. In fact, the capability to provide VoIP services such as Voice over LTE (while also supporting a great data throughput) is one of the key enablers for the LTE development. Particularly, VoLTE leverages telecom capabilities such as QoS and global grasp to deliver novel communication services, which clearly outperform OTT voice/video solutions such as Skype.

In contrast to earlier 3GPP wireless technologies, LTE does not rely on CS bearers to support voice. Therefore, the transmission of voice over LTE demands a migration to a VoIP solution (usually within an IMS environment). Furthermore, guaranteeing telco-grade voice involves the co-working of IMS/SIP with various LTE Radio Access Network (RAN) features. Ultimately, it is the combination of IMS/SIP with RAN features that provides a better user experience if compared to legacy CS voice services. The contribution of some RAN features to a carrier-grade VoLTE experience will be presented and discussed in this paper.

Table of Contents

1 Introduction

The increasing gap among capacity and demand poses an important call for novel network technologies to allow mobile operators to improve performance (from the end-users' point of view) on a cost-effective basis. In fact, Voice over Long Term Evolution (VoLTE) is a key component for an innovative set of base services defined for all-IP networks: the objective relies on making such new services as accessible as voice and SMS are nowadays, while also offering a flexible interaction with Internet applications.

Indeed, LTE focus on a (rather flat) all-IP access technology aiming at delivering a bandwidth-efficient method of carrying multiple types of user traffic at the same time. In other words, the capability of deploying Voice over IP (VoIP) services, while also supporting high-rate data throughputs, characterizes one of the critical drivers for the development to LTE.

Within this context, the IP Multimedia Subsystem (IMS) and Session Initiation Protocol (SIP) are essential technologies for deploying VoIP in an LTE setting. Nevertheless, the LTE RAN features are ultimately responsible for creating 'value added' services in relation to VoIP. Therefore, this paper illustrates in the second chapter the advantages of the IMS-based VoLTE approach as well as its (strategic) positioning compared to initial methods and over-the-top (OTT) providers.

Furthermore, the third chapter explains that optimization of radio features and parameters is required to offer reliable VoLTE connections based on high success rates. This is especially important for highly-competitive markets like the Austrian telecommunication sector: annual network performance tests (conducted by the P3 Group for example) may unmask a mobile operator's (network) vulnerabilities and critically influence its customer's preference and the likely churn rate. Finally, conclusions are drawn based on the key findings described in this paper.

2 Migrating to IMS-based VoLTE: Initial considerations

The need for bandwidth derived from devices and subscribers has been constantly increasing for many years. Actually, the data volume transferred by mobile networks is doubling approximately every year (Ericsson Mobility Report 2013) and the quantity of connected machine-to-machine (M2M) devices is estimated to exceed 50 billion by 2020. Still, one of the key performance indicators like the average revenue per user (ARPU) is decreasing in many markets.

As a result, to take advantage of mobile-broadband opportunities (while at the same time increasing profits/benefits for business and consumers) remains the focus of every operator's activities. In this sense, LTE networks are able to carry mobile broadband with huge data capacity and a minor latency level. Though, since there is no circuit-switched voice domain in LTE (Fig. 1), the telecom industry has implemented a universally interoperable IP-based voice and video calling solution for LTE: VoLTE, which further facilitates the evolution of innovative communication services.

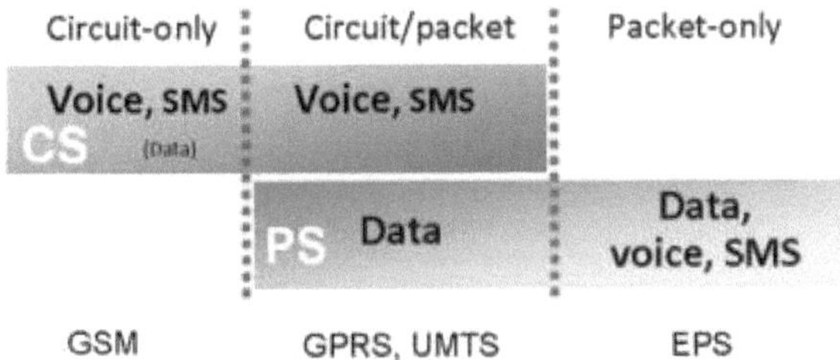

Figure 1: Circuit and packet domains. LTE within the Evolved Packet System (EPS) [1]

Furthermore, over-the-top (OTT) communication solutions like Skype (as well as other app providers) have influenced the way users assess a particular service based on VoIP. Nonetheless, a completely satisfactory user experience cannot be offered by OTT solutions due to missing QoS measures or the lack of handover mechanisms to the circuit-switched (CS) network. In addition, there is no guaranteed emergency support or extensive interoperability of services among diverse OTT services and devices. Thus, the readiness of subscribers to use a service that does not offer security, quality, flexibility or even mobile-broadband coverage clearly influence the adoption of OTT services in a negative manner.

2.1 Benefits of IMS-based VoLTE

On the contrary, VoLTE operators have an attractive competitive positioning: they can deliver mobile HD voice, implement video and messaging as well as converge with the web via Web Real Time Communications (WebRTC). Moreover, (VoLTE) operators can work in partnership with application providers and supply the best user experience on a low cost basis, since they can translate application developers' solutions onto innovative communication services. Besides, a quick customization of features for strategic activities (as the mobile healthcare industry could be) becomes feasible and effective.

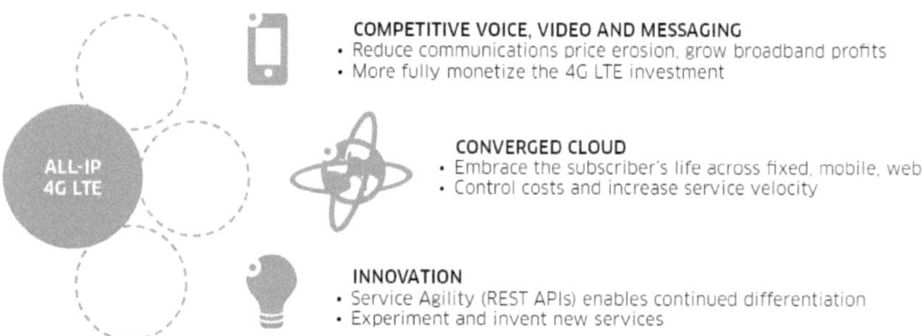

Figure 2: VoLTE's strategic value for operators [3]

Needless to say, VoLTE plays an important role in empowering all-IP communications in the 4G LTE network. As a result, VoLTE will enable operators to:

- Generate attractive communication services by merging mobile voice with video, the web and social networking
- Improve customer experience by delivering data and HD voice simultaneously (which also contributes to offload legacy infrastructure). In addition, the current 'fragmented' communications landscape that relies on various (rich) media approaches becomes more harmonized.

For sure, VoLTE is not the only method by which all operators will initially launch voice service in the short-run. Yet, it embodies the ideal approach if competitive advantages and their related (business) risks are considered.

2.2 Positioning VoLTE

Considering that there is no other well-specified, wide accepted or well-supported solution offered, operators will have no other choice than adopting VoLTE. This is true even if there are some provisional methods to convey voice in LTE devices (this is useful to operators aiming at selling out smartphones previous to the proper VoLTE's market maturity). As shown in Fig. 3, some approaches would usually rely on current circuit-switch 2G and 3G networks or, in a rather simple way, on app providers.

	SUBSCRIBER'S SERVICE	VoLTE NEW CONVERSATIONS	CSFB LEGACY VOICE & SMS	APP PROVIDERS MANY SERVICES
Standards	End to End QoS	✓	✓	✗
	Global interoperability, including regulatory	✓	✓	✗
	Roam with local voice, not home-routed data	✓	✓	✗
Multimedia	All-IP network enables video-comms, etc.	✓	✗	✓
	4G LTE data simultaneous with voice	✓	✗	✓
	Foundation for services innovation, WebRTC, etc.	✓	✗	✓
Voice	Evolved voice: HD, new features, WebRTC, etc.	✓	✗	✓
	Minimal voice call setup delay	✓	✗	✓
	Graceful continuity to 2G/3G circuit voice	✓	✓	✗

Figure 3: VoLTE compared to other preliminary approaches for voice in LTE devices [3]

Particularly, the Circuit-Switch Fallback (CSFB) and the dual-radio methods on which CDMA operators rely (in some cases known as Simultaneous Voice-and-LTE, SVLTE) are convenient to the extent that current telephony services are reused. However, these are not indeed LTE solutions and, as a result, are not subject to suitable development. Likewise, OTT methods are attractive since they run over IP and support, on the one hand, richer experiences. On the other hand, important features like interoperability and 2G/3G integration/handover are (with the OTT approach) definitely at stake.

Furthermore, CSFB voice is characterized by the limited service since a voice call (originating or terminating) leads to a fallback to legacy 2G/3G service. This implies the restriction to CS voice to the subscriber, including slower 3G mobile data services or its complete loss during a fallback to 2G. Nevertheless, operators who temporarily choose CSFB should critically reflect on the invested resources due to:

- Communication experience for end-users might be at stake: during voice calls CSFB subscribers would be downgraded from 4G LTE data service to 3G HSPA+ or 2G, losing data services completely.
- Boundaries are imposed on innovation: New services that work on all-IP networks cannot be realized (such as video calling)
- Increased OPEX and risk: the economic and technical challenges that rise with the operation and maintenance of 4G LTE and legacy 2G/3G networks in parallel should not be underestimated.

Despite the fact that HD voice is being gradually implemented in 3G networks, 4G LTE networks would be more present in the years to come along with more capable devices and network elements. In terms of call setup time, CSFB requires 4 seconds (around 1 additional second compared to the 3G's required time) and approximately twice the setup time necessary for a VoLTE's call (idle to idle state, see [5]).

3 VoLTE Parameter Optimization

Deploying VoLTE requires a number of optimization steps to gain the full benefit of the technology's potential. Ideally, the success rate and the retainability of a VoLTE call must exceed the level provided by CS connections.

Basically, network optimization involves the activation of features and optimization of parameters such as:

- Robust Header Compression reduces the bandwidth associated with the headers used to transport relatively small encoded audio packets
- Transmission Time Interval (TTI) Bundling overcomes the limitation of using short (1ms) TTIs at cell boundaries

These features contribute to make a VoLTE call reliable while still providing more efficiency compared to OTT VoIP applications. Special attention is paid, therefore, to decreasing the necessary bandwidth for voice and get the most out of the capacity. By doing so, the handover, setup and call completion success rates could be further improved.

In terms of handset power consumption, the focus lies on the one hand on the enhancement of the handset architecture: the integration of VoLTE into the chip set or the sleep mode available to the application processor fall into this category. On the other side, radio features like the Discontinuous reception (DRX) intended to conserve the UE's battery life during a VoLTE conversation complement the (handset) power saving solutions. Other valuable drivers are:

- Dedicated Bearers allow for the prioritization of VoLTE audio packets over all other best-effort traffic
- Semi-Persistent Scheduling (SPS) reduces the complexity and overhead of the continuous allocation of DL/UL physical layer resource blocks to transport the audio traffic

Additional (optimization) features in the field of speech quality, depending primarily on the voice codec sampling rate and the derived audio bandwidth, will be discussed separately.

3.1 Robust Header Compression (RoHC)

RoHC takes advantages of the redundancy of information present in

- the headers of subsequent packets in the same audio stream and
- various headers in different protocol layers

with the goal of decreasing the header size used to transport VoLTE audio. In fact, the 40-60 bytes of header length can be condensed to 3-4 bytes. And this applies to VoLTE calls, which usually comprise small encoded audio packets transmitted every 20ms.

Particularly, the size of the data is smaller than the headers for the protocols used to carry the encoded data. For VoLTE deployments based on IPv6, the mix of RTP/UDP/IP headers can sum up to 40-60 bytes long headers. This means that a VoLTE encoded audio transmission (relying on the Wideband-AMR codec and RoHC) is reduced from 75 bytes to approximately 35 bytes.

As shown in Fig. 4, RoHC is used over the air interface to preserve the RAN's valuable bandwidth. In fact, RoHC may accomplish around 50% reduction in the size of VoLTE audio transmissions: the required bandwidth for any call is reduced while the amount of users on a given eNodeB site is increased.

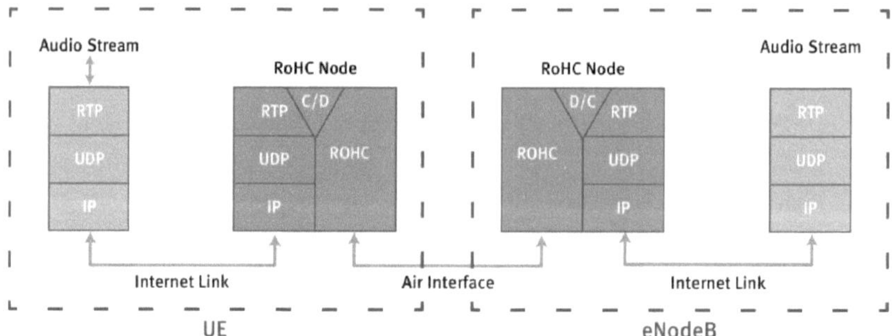

Figure 4: RoHC Compression and Decompression at the UE and eNodeB [6]

3.2 Transmission Time Interval (TTI) Bundling

LTE presents a shorter TTI (1ms subframe) than earlier defined in cellular technologies. Clearly, a smaller TTI enables low over-the-air latency for real-time applications as resource scheduling is performed for every TTI. Some uplink issues are, though, caused by the short TTI in particular settings of an eNodeB's coverage: given a UE located at a cell edge (with deteriorating reception and no possibility of enhancing its transmission power), the eNodeB may activate the TTI bundling. In other words, the UE will intensify the error detection and correction related to each data broadcast by transmitting over several TTIs. Based on this improved approach (e.g. error detection/correction), overall latency is diminished compared to the use of a single TTI.

As presented in Fig. 5, TTI bundling contributes to lower-latency VoLTE data at cell edges, where information errors are predictable. Instead of waiting for the HARQ process (typically 8ms/period) to send a data retransmission request, a data retransmission is assumed by TTI bundling: by doing so, several data packets are filled into a HARQ interlace period in advance. As a result, every packet comprises the identical source data coded with 4 distinct groups of error detection/correction bits.

Generally speaking, the short 1ms TTI (along with its bundling approach) represents for VoLTE a considerable advantage: it improves the uplink performance at cell edges by means of multiple bundled TTIs to transfer higher error detection and correction information.

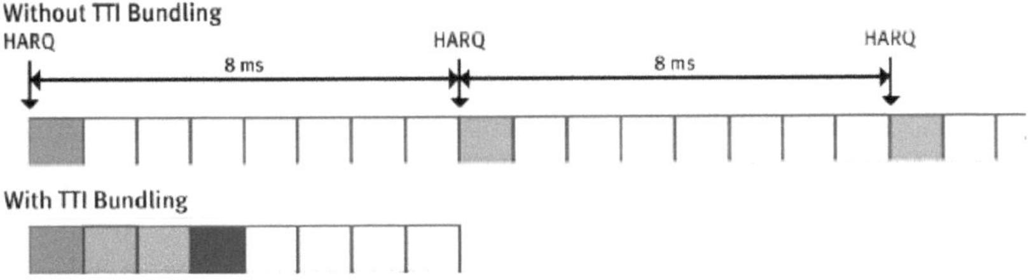

Figure 5: Effect of TTI bundling on latency [6]

3.3 Discontinuous Reception (DRX)

For packet-based voice services, each encoded audio packet transmission (usually 20ms for VoLTE) is followed by a period of no transmission. DRX exploits, for instance, these silent periods by switching off several components such as digital signal processors, UE's RF receiver or A/D converters. As a result, the device's battery life can be conserved.

Nevertheless, setting a too long 'sleep period' might involve latency issues at the detriment of the wanted QCI value level (e.g. a predefined performance profile). Particularly, the network establishing the DRX pattern (and also being aware of the data downlink schedule to the UE) should carefully select a suitable DRX configuration in view of likely retransmission needs and latency constraints of the application. In this sense, special attention should be paid to the influence of TTI on DRX in LTE so as to improve DRX settings in favor of power saving.

As depicted in Fig. 6, the two DRX operation modes can be applied according to the pause duration within a chat: a long DRX deactivates the UE receiver for a longer time period in a situation where audio packets are not frequently delivered. The eNodeB's MAC Layer or an activity timer at the UE would regulate the transition from short to long DRX and vice versa. Worth mentioning is the absence of DRX short or long cycles in UMTS (only fixed-length ones are available).

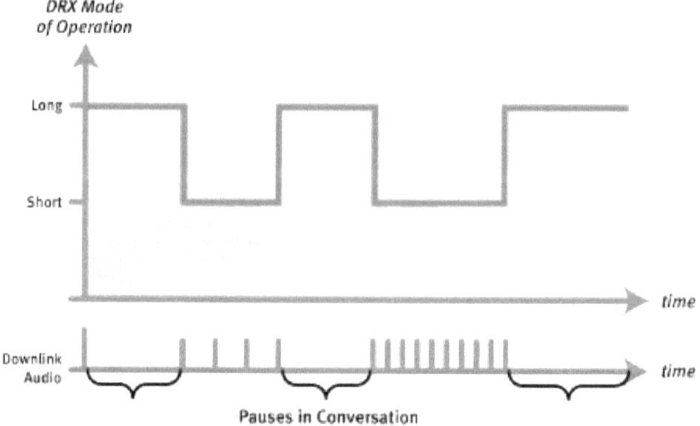

Figure 6: Long and Short DRX [6]

3.4 Dedicated Bearers

Despite the increased spectral efficiency offered by LTE, over-the-air bandwidth remains a limited and valuable resource. Moreover, Internet-based VoIP services usually generate a greater load on mobile networks, leading to negative (quality) side-effects. Operators need, therefore, to invest in more capacity to cope with the OTT voice traffic since (their) voice applications naturally consume additional network resources. In other words, each application along with the associated data strives for that finite bandwidth.

In fact, the encoded voice packets an off-the-shelf VoIP client produces do not show a significant difference when compared with data traffic derived from web browsing or video streaming (from a network's perspective). As a result, the network tries to bundle all users' (generic) data traffic into a single common channel[1]. Interestingly, with a default EPS Bearer no control over the service quality is possible. This implies a 'best effort' approach to transport all generic traffic to the Internet PDN. By the time the network resources are (temporarily) exhausted, latency may greatly fluctuate and packets be dropped due to data traffic queuing.

To overcome these drawbacks, which cannot be tolerate for real-time applications, LTE introduces the EPS Dedicated Bearer: It permits the isolation of certain types of data traffic (VoIP traffic versus FTP file download for instance). For VoLTE, an EPS dedicated Bearer will exclusively convey encoded voice packets among the UE and an IMS PDN-GW. Also, each Dedicated Bearer can possess diverse quality characteristics defining a QoS Class Identifier (QCI).

- Packet Delay Budget: The Maximum tolerable end-to-end delay among the UE and the PDN-GW
- Resource Type: Guaranteed Bit Rate (GBR) and Non-GBR
- Allocation Retention Priority: In case the capacity is depleted, scheduling is based on the value assigned ('1' equals to the highest level)
- Packet Error Loss Rate: Threshold level of IP packets not effectively received by the Packet Data Convergence Protocol (PDCP)

---

[1] e.g. the Physical Downlink and Physical Uplink Shared Channels PDSCH/PUSCH, having one or more EPS bearers linking the UE to the PDN-GW

Figure 7 further presents standardized QCI values, '1' being assigned to VoLTE traffic that would be prioritized over all best-effort traffic on the Default Bearer.

QCI	Resource Type	Priority	Packet Delay Budget (ms)	Packet Error Loss Rate	Example Services
1	GBR	2	100	10^{-2}	Conversational Voice
2	GBR	4	150	10^{-3}	Conversational Video (live streaming)
3	GBR	5	300	10^{-6}	Non-conversational video (buffered streaming)
4	GBR	3	50	10^{-3}	Real-time gaming
5	Non- GBR	1	100	10^{-6}	IMS Signalling
6	Non- GBR	7	100	10^{-3}	Voice, video (live streaming), interactive gaming
7	Non- GBR	6	300	10^{-6}	Video (buffered streaming)
8	Non- GBR	8	300	10^{-6}	Video (buffered streaming)TCP-based (email, www, FTP)
9	Non- GBR	9	300	10^{-6}	

Figure 7: Standardized QCI Values. Adapted from [6]

3.5 Semi-Persistent Scheduling (SPS)

As previously stated, the physical layer relies on common channels (PDSCH/PUSCH) to deliver the data contained in the logical bearers. A way to allocate these channels appears necessary to avoid several users (on an eNodeB) attempting to use identical resources at the same time. Specifically, a LTE carrier is split into several subcarriers within the frequency domain, whereas in the time domain every subcarrier is gathered into 0.5ms time slots.

A cluster of twelve subcarriers in one time slot is referred to as a Resource Block (RB) and is defined as the smallest portion of the LTE physical layer (resource) that can be placed to a UE (refer to Fig. 8). Moreover, VoLTE encounters a challenge when it comes to granting control channel overhead: as each downlink/uplink RB must be granted, the (control channel) overhead for the ongoing allocation of RBs becomes unmanageable.

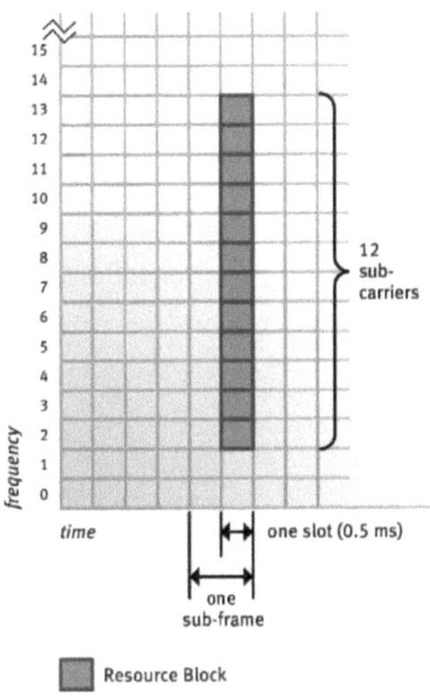

Figure 8: LTE Resource Block (RB) in the time and frequency domain (Physical Layer) [6]

To overcome this, Semi-Persistent Scheduling (SPS) was introduced to reduce granting overhead for applications such as VoLTE. In effect, the steady transmission configuration of VoLTE packets allows SPS to make a continuing grant of RBs instead of separately arranging each download/uplink RB. Of course, this involves a Radio Resource Control (RRC) message used to determine the RB grant's periodicity.

The green boxes in Fig. 9 illustrate the SPS-arranged RBs for a VoLTE call. As further shown by the orange box, additional RBs can be dynamically organized for data traffic while SPS is enabled (for example for a file download during a VoLTE call).

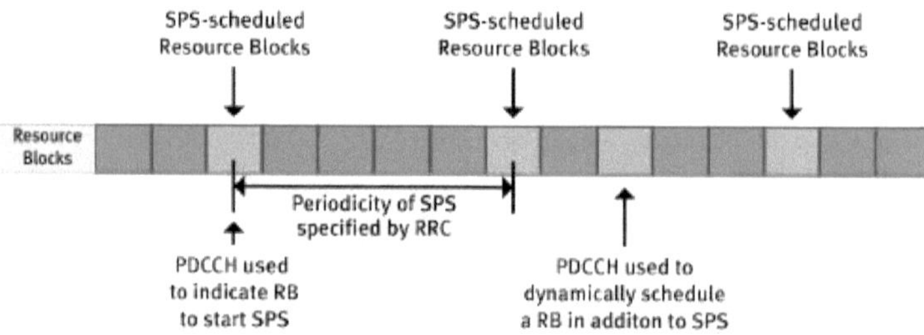

Figure 9: Semi-Persistent Scheduling [6]

Important to consider are the pauses (e.g. short-time silence periods) in the course of a VoLTE conversation: this may represent a drawback since (physical layer) resources get lost in case the SPS grant is continued.   In this case, the expiration of the SPS grant after some network-defined transmission slots would be recommendable (uplink). In the end, it's all about finding the right balance between efficiency improvement (for shared data channels) and minimizing control channel overhead.

4 Conclusion

VoLTE may boost an operator's network productivity and further enrich the subscriber's experience in terms of call setup times and speech quality. Consequently, VoLTE embodies the operator's imminent strategy and progress towards novel communication services this industry must deal with. For instance, it is not a matter of deciding whether to implement VoLTE or not, but instead of selecting the proper tactical approach to do it.

Furthermore, major efforts should be undertaken to deal with events threatening a first-class voice experience, e.g. being outside LTE coverage (prior to a conversation) or quitting the LTE coverage range after a call origination. This scenario is indeed relevant since extensive LTE coverage, throughout the initial LTE deployments, cannot be assumed.

Meanwhile, IP communications are being accepted as the next evolutionary step of core mobile services. Operators are, therefore, called to continue investing in their networks (e.g. IP infrastructure) so as to enjoy a superior brand positioning and be able to 'switch-off' legacy network in the medium term. Additionally, greater control over supplementary access methods (Wi-Fi for example) as well as an increased need for operator's core applications is the VoLTE promise in the long run.

Looking ahead

Despite some deployment issues or extensive consensus, the telecom sector gives a good impression of being prepared for IP-communications. In this regard, the 'journey' away from industry-driven towards to demand driven growth is expected as a direct consequence of the emergent acceptance of HD calling and VoLTE in the subscriber's mind [30]. Undoubtedly, VoLTE is gradually being perceived as a premium device's in-built feature, likely to be found already 'in the box'.

Finally, no matter where competitive pressures or regulation initiatives of technology enhancements might take the telecom business within the next few years, operators' (financial) endeavors in favor of VoLTE will enable them to handle in a forward-looking and customer-friendly way.

Bibliography

[1]   3GPP, LTE, available on http://www.3gpp.org/technologies/technologies [last access 10.03.2015]

[2]   Aalborg University, Nokia Siemens Networks, Power Savings and QoS Impact for VoIP Application with DRX / DTX Feature in LTE, 2011

[3]   Alcatel-Lucent, Voice over LTE: The new mobile voice, Strategic White Paper, 2014

[4]   Ericsson, Voice and video calling over LTE, White paper (Rev B), 2014

[5]   Nokia Networks, Voice over LTE (VoLTE) Optimization, White paper, 2014

[6]   Spirent, VoLTE Deployment and the Radio Access Network. The LTE User Equipment Perspective, White paper, 2012

[7]   Signals Research Group, Behind the VoLTE curtain (Part 1), Vol. 10, Number 7 PREVIEW, 2014

[8]   Swisscom, Annual Report 2014, available on https://www.swisscom.ch/en/about/investors/reports.html [last access on 15.03.2015]

[9]   V. Paisal (Samsung India Software Operation Pvt Limited), Seamless Voice over LTE, 2010

[10]   Weka Media Publishing, Anruf der vierten Generation, Connect Magazine 12/2014, pages 70-74

List of Figures

List of Abbreviations

3GPP	3rd Generation Partnership Project
AMR	Adaptive Multirate
ARPU	Average Revenue Per User
DL	Downlink
eNB	Evolved NodeB
HSDPA	High Speed Downlink Packet Access
IEEE	Institute of Electrical and Electronics Engineers
IP	Internet Protocol
LTE	Long Term Evolution
OTT	Over-The-Top
QoS	Quality of Service
UE	User Equipment
TTI	Transmission Time Interval
UMTS	Universal Mobile Telecommunications System
UTRAN	UMTS Terrestrical Radio Access Network
VoLTE	Voice over LTE
VoIP	Voice over IP